D1180769

Artists' Kew

SERGEI SUKONKIN
A Walk Along the Embankment
watercolour
29 x 41 cm

An exhibition of contemporary works

9 May–18 June 2006
Kew Gardens Gallery

25–27 April 2006
Preview at Messum's
Cork Street, London W1

Under the auspices of the Asgill House Trust
in association with the Foundation and Friends of the
Royal Botanic Gardens, Kew (registered charity no. 803428)

DAVID MACH

A & E
collage on board
10 x 7 cm

First published in 2006 by
Royal Botanic Gardens, Kew
Richmond, Surrey, TW9 3AB, UK
www.kew.org

ISBN 1 84246 143 5

British Library Cataloguing in Publication Data
A catalogue record for this book is available
from the British Library

Design by Culver Design
Colour reproduction by Saxon
Photolitho Ltd, Norwich
Printed and bound in Italy by
Mondadori Printing

For information or to purchase
all Kew titles please visit
www.kewbooks.com or email publishing@kew.org

Patrons

Lord and Lady Attenborough of Richmond upon Thames
The Earl and Countess of Selborne
Baron and Baroness van Dedem

Baroness van Dedem
Chairman

David Messum
Fine Art Consultant

Graham Ball
Financial Consultant

Jane Faust
Media Consultant

Fred Hauptfuhrer
Exhibition Coordinator

For our volunteer group, *Artists' Kew* follows on a similar exhibition we organised three years ago. With the celebrated Richmond Hill setting as its theme, that exhibition was called *The View*, and it raised more than £100,000 for local Conservation Area projects.

Both exhibitions were the idea of Fred Hauptfuhrer, whose home in Richmond is Asgill House. Our adjoining riverside properties are on the site of the former Tudor Palace there. So I suppose it is only natural that we share a special interest in the historical associations between Kew and Richmond.

We were encouraged from the start by the enthusiasm and advice of Professor Sir Peter Crane FRS, Director of Kew Gardens, and Lucy Blythe and Amy Stockwell at the Foundation & Friends of Kew. It has been a pleasure throughout to work with them and their dedicated staff. Sir Peter, alas, is leaving Kew shortly after our exhibition to take up a professorship at the University of Chicago (he had been at the Field Museum in Chicago for 17 years before becoming Kew's Director in 1999). We wish him well, and shall miss him.

We are, above all, indebted to the contributing artists who have made the exhibition what it is, with added thanks to those offering an extra portion of their share of the proceeds to the exhibition's charitable purpose.

Ronny van Dedem
Richmond

February 2006

Baroness van Dedem, Chairman of The View *as well as* Artists' Kew, *is herself an artist (specialising in portraiture) and sculptor. She lives at Trumpeters House with her husband Willem, who is President of the European Fine Art Fair held annually in Maastricht.*

A View of the South Side of the Ruins at Kew, engraved by William Woollett (1735-1785) after a drawing by Joshua Kirby (1716-1774), from William Chambers (1723-1796) *Plans, Elevations, Sections and Perspective Views of Gardens and Buildings at Kew,* 1763

Foreword

It is a striking fact that the Royal Botanic Gardens has attracted artists, apart from botanical ones, so very rarely. Here, we would have thought, was a subject in close proximity to the heart of the metropolis which should have been a mecca for painters. Not so. Painters like the Impressionists would have loved to record Kew and its visitors. Pissarro did. But think what Renoir and others would have made of it! Instead we have a burst of activity in the middle of the eighteenth century and then it fizzles out. But those early works remain formidable ones to emulate. I think of Richard Wilson's atmospheric view towards the pagoda and bridge and his rendering of the ruined classical arch built by Sir William Chambers. Almost immediately after, in 1763, came the famous series of etchings by Paul Sandby published in Chambers' folio volume dedicated to the garden. And that, more or less, is it.

It is, therefore, an inspired idea to lure painters back to Kew after so long a time lapse. And for what better cause to lure them than an exhibition the proceeds of which will go towards a new gallery in which the Royal Botanic Gardens' legendary collections of botanic art will be displayed. In this event too a flag is raised signalling the imminence of the Gardens' 250th anniversary.

Roy Strong
Much Birch, Herefordshire
February 2006

Sir Roy Strong, writer, historian, diarist and gardener, was Director of the National Portrait Gallery, 1967-73, and the Victoria and Albert Museum, 1974-87, where he mounted a series of seminal exhibitions on the future of the historic environment.

Frederick, Prince of Wales and his sisters making music in the gardens of the White House, with the Dutch House in the background. Oil painting by Philip Mercier, 1733
Reproduced with permission of the National Portrait Gallery, London

Introduction

To the world at large, 'Kew' means the Royal Botanic Gardens - and indeed on 3rd July 2003 Kew Gardens was given the status of a World Heritage Site, celebrating the beautiful grounds and the unique collection of plants and Kew's enormous importance as a scientific institution.

Kew is more than the gardens, though. Kew, today one of London's loveliest suburbs, is an ancient community, centred round its Green, where princesses and potentates lived five centuries ago, where some 250 years ago a Prince of Wales and his wife conceived the idea of devoting part of their garden to exotic plants, where artists lived and musicians serenaded the Royal Family, and where now hundreds of thousands of visitors come to visit the Gardens or to research at The National Archives and where thousands of fortunate people can count themselves at home.

This exhibition makes the point. The artists taking part have varied their contributions between the Gardens and the wider Kew community.

Kew first developed as a tiny hamlet on the Surrey bank of an important ford across the River Thames. This ford may have been where Julius Caesar's legions crossed the river or where the Emperor Claudius's army crossed a century later, putting the Britons to flight at the sight of armoured war elephants. It was certainly where King Edmund Ironside's army crossed to defeat the Danes under Cnut in 1016. The river itself was important not only as a conduit to London, but in the evolution of Kew's royal history as the link among royal palaces from Hampton to Greenwich.

We first have any description of Kew in 1314, in a survey of the manor of Shene (later Richmond). There were just five manorial tenants living at Kew with their families. It was where much of the woodland of the manor was found, and the manorial meadows stretched along the riverbank. Part of these meadows and a fishery in the river belonged to the Priory of Merton; another fishery which included the pond at the corner of the Green belonged to the monks of St. Swithin's (the Cathedral) in Winchester. The Green was much larger than now and had a long narrow 'panhandle' stretching out for some 500 yards at its west end, almost to the point where the ford had been replaced by a ferry.

It was in the time of the Tudors that Kew first became a fashionable place to live. Conveniently close to, but equally conveniently just far enough away from, the royal palace of Richmond (so renamed by Henry VII when he rebuilt the former palace of Shene), Kew became the home of grandees. Among them were the Earls of Devon, Princess Mary Tudor and her husband the Duke of Suffolk, King Henry's cousin the Earl of Worcester, the Earl of Hertford (later Duke of Somerset, brother of Queen Jane Seymour), John Dudley (later Duke of Northumberland) and then his son Robert Dudley, Queen Elizabeth's favourite as Earl of Leicester. In the reign of James I, it was home to Princess Elizabeth Stuart, ancestress of the Hanoverian kings of England and so of our present Queen.

The original cottages stood by the riverside, spread out from where now is the Brentford Gate of Kew Gardens about as far as the Herbarium on the north-western side of the Green. From the late 16th century onwards, plots of land from both north and south sides of the Green were progressively granted out for the construction of new houses and cottages. By the end of the 18th century, nearly all of the perimeter of the Green had been developed and one or two houses had even been built along the roads leading to Richmond and Mortlake. But in 1801, Kew still had only 85 houses and a total population of 424 persons.

Among those houses there were, however, royal palaces. James I had made his Richmond Park on the north-west side of Richmond Palace, stretching up on the Surrey side of the river about as far as today's lake in Kew Gardens and extending eastwards from the river as far as Holly Walk. In the centre of his park, James built a hunting lodge. After the destruction of Richmond Palace, this lodge was enlarged by King William III and then for the Duke of Ormonde. In 1718, the new Prince of Wales (George II from 1727) and his wife Caroline acquired it as a country house. Caroline had the northern part of the grounds laid out as 'Richmond Gardens' by Charles Bridgeman, with a riverside terrace and embellished with such follies as 'Merlin's Cave' and a 'Hermitage' designed for her by William Kent. The southern part of the gardens was left as the 'Old Deer Park'. As Richmond Lodge was too small to house an expanding household, in 1728-30 the Queen rented more ground and extra houses by the Kew riverside from Robert Thoroton and his wife Mary, daughter of Richard Levett, a former Lord Mayor of London. These properties included the 'Dutch House' (today's Kew Palace), which had been built by a merchant in 1631.

Cattle grazing in the Old Deer Park. Watercolour by W H Fitch (1817-1892)

Just a year after the completion of these new leases, Frederick Prince of Wales moved in next door to his parents. To his father's annoyance, he acquired a lease of the house with 20-acre grounds which stood opposite the Dutch House across the narrow tip of Kew Green. No more than 50 feet separated the two properties, and a mere 200 feet the two buildings. Frederick at once got William Kent to re-model the house. It was now plastered on the outside and became known as the 'White House' (since demolished) - in distinction to the red brick Dutch House.

Frederick then enlarged his grounds by purchasing and renting extra fields along 'Love Lane', which bordered the east side of Richmond Gardens, until his property extended alongside that of his parents as far south as the drive linking Richmond Lodge to the Kew Road. He and his wife Augusta planned elaborate new gardens - and after Frederick's untimely death in 1751, Augusta carried through these plans. They included the creation of a small physic garden for exotic plants as well as a score of garden buildings designed by William Chambers, of which the main survivors today are the Orangery, the Pagoda and the Ruined Arch. The year 2009 will be celebrated as the 250th anniversary of the botanic gardens.

In these projects, Augusta had the help and advice of her close friend John Stuart, 3rd Earl of Bute, who also had a great influence over her eldest son the young Prince George, and fired the latter's interest in botany. When George II died in 1760 the young prince, now George III, inherited Richmond Lodge and Gardens (which he hired Capability Brown to re-

George III and Queen Charlotte walking in Kew Gardens. Hand-coloured engraving, 1787. *Reproduced with permission of the Museum of Richmond*

develop) and the lease of the Dutch House. Then on Augusta's death in 1772 George also inherited the lease of the White House and Kew Gardens. He and his wife Charlotte then moved their country home to the White House from Richmond Lodge (which was demolished). With the help first of Bute and then of Sir Joseph Banks (unofficial director of the gardens from the early 1770s) they continued to develop and expand the exotic 'botanic' aspect of the gardens. It was at Kew that George recovered from his first serious bout of porphyria, his 'madness', in 1788-9. In 1785 he had been able to close 'Love Lane' separating his two estates, and when in 1802 he had finally been able to buy the freehold of the Kew estate from the heirs of the recently deceased Earl of Essex the last fences between Richmond Gardens and Kew Gardens were removed.

View of the Castellated Palace from the Kew Palace lawn. Artist unknown, c.1810-1820

George III's purchase of the Kew estate also enabled him to proceed with a pet project for building a new palace. (He had already started one designed by Chambers to be built near Richmond Lodge, but this was abandoned after the move to Kew and cleared away in the 1780s.) This time, he commissioned James Wyatt to build him a Gothic castle, the 'Castellated Palace' by the riverside. The White House was demolished and the Royal Family moved into the Dutch House as a pied-a-terre while awaiting the completion of the new building. But the King's ultimate lapse into madness in 1810 put a final stop to the work, with the exterior completed but the interior not yet finished. It was never occupied. Queen Charlotte died in the Dutch House, by then known as 'Kew Palace', in 1818, and in 1827 George IV gladly agreed to the demolition of the Castellated Palace. He had plans himself for a new building at Kew, for which he enclosed the western end of the Green, but had to abandon them for lack of funds.

Shortly after her accession Queen Victoria agreed that control of the gardens should be transferred to the Commissioners of Woods and Forests, to be developed as a national botanic garden. This was done in 1840, and in 1841 Sir William Hooker was appointed as the first official Director of the Royal Botanic Gardens. Starting with little more than the original botanic garden, he was gradually given more and more of the former Kew and Richmond Gardens, though the Queen kept for herself Queen Charlotte's Cottage and its surrounding woodland, as well as Kew Palace. Under Hooker and his son Joseph who succeeded him as Director, the Gardens were re-designed and great new plant houses including the Palm House (1844-48) and part of the Temperate House (1859-1899) were built. The scientific side of the gardens was developed, and with it grew Kew's wealth of art, which was collected originally for its scientific value. Queen Victoria celebrated her Diamond Jubilee in 1897 by presenting Queen Charlotte's Cottage and its surrounding grounds to the Gardens, and by arranging that Kew Palace should be opened to the public.

To this day Kew Gardens continues to meet the conditions of Queen Victoria's gift. In contrast to the glasshouses and formal planted areas, the Gardens are obliged to maintain the grounds of Queen Charlotte's Cottage in their natural state. Today they form an important conservation area supporting biodiversity and protecting indigenous plant species, while Kew Palace is restored to its former glory following a long period of closure and major restoration by Historic Royal Palaces.

The presence of the Royal Family at Kew in the late 18th century had given a boost to the village's development. The building of a wooden bridge across the Thames connecting Kew and Brentford in 1758-9 had led to the improvement of the main road to Richmond (and then to the closure of Love Lane). The first bridge was not very satisfactory, and it was replaced by a new stone bridge of nine arches, designed by James Paine, in 1783-9. This was in turn replaced by the present bridge in 1903.

The gentlemen of Kew, tired of having to make the journey to Richmond every Sunday in order to go to church, had in 1710 got up a subscription to build a chapel, and they persuaded Queen Anne to grant a plot of land on Kew Green for its site.

View of St. Anne's Church on Kew Green. Sepia drawing. Artist unknown, 1785

The new little chapel, which had the status of a 'chapel of ease to the parish of Kingston within the chapelry of Richmond', was dedicated to Saint Anne in May 1714. In 1769 Kew was linked with Petersham in a new parish and vicarage, and the chapel became a parish church. Enlargement of it was carried out by Joshua Kirby, the Royal Clerk of Works, and was funded by George III himself. It has been further enlarged several times since.

In its churchyard lie Kirby and his friend Thomas Gainsborough (who rented a house on the Green for a few years in the 1780s). The artist John Zoffany who lived across the river at Strand on the Green in Chiswick is also buried there. Other famous painters who lived for a while at Kew were Sir Peter Lely, the portraitist of Charles II's court; Jeremiah Meyer, miniaturist to George III; George and John Cox Engleheart, also miniaturists; Walter Deverell and Arthur Hughes, both pre-Raphaelites; and Camille Pissarro and his son Lucien. A number of artists in this exhibition also happen to live in the immediate area.

Many of the attractive houses round Kew Green date from the 18th and early 19th centuries. A few have splendid plaster ceilings by Francis, the founder of the Engleheart family in England, and his son John Dillman Engleheart. (Francis had married the niece of John Dillman, head gardener to Prince Frederick and Princess Augusta.) The last houses to enclose Kew Green were built in its northeast corner in the early 19th century. They included Kew's first purpose built school, designed in the Gothic style, standing on land overlooking the pond provided by Miss Elizabeth Doughty, who had a private house designed as a Gothic priory in the large grounds behind the east side of the Green. King George IV contributed handsomely to the new school, and when it was opened in 1824 decreed that it should be called 'The King's Free School'. It has since changed from King's to Queen's and back according to the sex of the monarch. The Gothic schoolhouse was replaced in 1887 by a new building, which was in turn demolished and replaced by a row of town houses after the school moved to new premises in Cumberland Road in 1970.

Also round the Green are some of Kew's public houses. A number of small alehouses are traceable to the early 18th century. By the end of the century three had survived as substantial inns or taverns. All but the Coach and Horses were re-built in mock-Tudor styles in the early 20th century.

The development which sparked the most dramatic growth of Kew was the construction of what is now the District underground line in 1869 and the opening of a new Kew Gardens station. Very quickly the land between the Kew Road and the Mortlake boundary became filled with new roads and houses. The station had its own buffet, a feature then unusual for a small suburban station. There was no refreshment facility in the Gardens until 1888, so this was a useful commercial venture. It later became the Pig and Whistle public house and is now the Flower and Firkin.

As visitors to the Gardens grew in numbers, by the early 20th century almost every house on the north side of the Green was either a pub, a restaurant or 'tea gardens'. As better facilities were provided in the Gardens, these gradually closed down, but a tea shop that survived was Newens bakery, established in Kew Road, near the junction of Kew Gardens Road leading to the station, in the 1880s. In 1957 Newens (re-built after war time bomb damage) bought the name, goodwill and secret recipe for the cheese-flavoured tartlets known as 'Maids of Honour' which had been sold for 250 years at a shop in Hill Street, Richmond. So visitors to the Gardens can now enjoy 'original Maids of Honour' on the way back to the station.

In 1890 Richmond was incorporated as a borough. Two years later, it absorbed not only Kew and Petersham but a slice of Mortlake on the east side of the railway. Originally market gardens, a small part of the area close to Kew Gardens station was built up at the beginning of the 20th century, but most of what is now called North Sheen was not developed until the 1920s and '30s. It was provided with a church by the removal of a large old barn from Oxted in Surrey. Re-built in North Sheen, the 'barn church' was dedicated to Saint Philip and All Saints in February 1929.

In the 1920s a large single-storey complex was developed on land near the river to the east of the District Line tracks to house the Ministry of Labour Claims and Records Offices. That site was subsequently re-developed to become the new Public Record Office, recently renamed The National Archives.

It is remarkable that Kew is now home to both The National Archives and the Royal Botanic Gardens. The National Archives contain some of the world's greatest documentary treasures ranging from the Domesday Book to original writings by influential politicians, scholars and artists. From these treasures it is only a short walk along the river to the largest and most important collections of living and preserved plants anywhere in the world, a vast and distinguished botanical library, and scientific research programmes that inform the conservation of plants across the globe.

As a backdrop to the varied art in this book, the visual representation of plants has been central to Kew's work since the late 18th century. What began with a scientific raison d'etre took on its first important dimension in 1790 when Sir Joseph Banks encouraged and supported the young artist Francis Bauer to come and live in Kew: in a house that still stands on Kew Green. His task was to record the remarkable influx of plants to the royal estate from all over the world. It was Bauer who produced some of the first illustrations of the bird-of-paradise flower brought back from the Cape by Francis Masson, one of Kew's first plant collectors. And it was Bauer, along with his contemporary Margaret Meen, who instructed the women of George III's family in the art of botanical illustration.

In the 19th century Walter Hood Fitch and his contemporaries masterfully captured the essence of Kew's new intake of exotics: the giant waterlily Victoria from South America, bulbs from the drylands of southern Africa, and multitudes of newly discovered orchids from the forests of south east Asia. At the same time, Marianne North travelled to all continents except Antarctica in search of the scenes and plant subjects exhibited in the gallery which carries her name.

Over the past century the tradition has carried on: an unbroken thread connecting Bauer and Meen with the exquisite artistry of Lilian Snelling in the early 20th century and that of Margaret Mee, Margaret Stone, Stella Ross-Craig and Mary Grierson which came later.

Many of these and other artists lived in the community, as do some of their worthy successors today. As well, aspiring artists come to Kew to study the collection of botanical art and to train - as Queen Charlotte did - with the experts. They doubtless will help Kew to continue nurturing the synergy of Nature and Art so vividly expressed in the pages which follow.

John Cloake
Richmond
February 2006

Bird of Paradise Flower (Strelitzia reginae). Hand-coloured lithograph by Francis Bauer (1758-1840) from his monograph *Strelitzia Depicta*, 1818

John Cloake, a retired diplomat and founding father of the Museum of Richmond, is the Borough's most respected local historian. He is President of the Richmond Local History Society and has written seven books and many monographs and articles about the history of Richmond and Kew.

Artists' Kew

OLGA GEOGHEGAN

The Orangery at Night
watercolour
29 x 40 cm

Late Afternoon Stroll
oil on board
20 x 30 cm

Evening Sky, Kew Bridge
oil on board
30 x 41 cm

The Temple of Bellona and the Ghost of Kew Gardens
oil
91 x 61 cm

The Orangery
oil
63.5 x 84 cm

Flower Bed
pastel
95 x 95 cm

Kew Bridge at Low Tide
watercolour
70 x 100 cm

A View from Kew
watercolour
70 x 100 cm

Hot House Secrets
watercolour
58 x 65 cm

Bloomin' Kew
watercolour
58 x 65 cm

In the Palm House
watercolour
28 x 39 cm

Admiring the Fish in the Palm House
watercolour
28 x 39 cm

A Quiet Drink in Browns Restaurant
watercolour
39 x 28 cm

The Boy in Kew Gardens
watercolour
27.5 x 25.5 cm

The Siesta
oil on canvas
101 x 91 cm

The Family
oil on canvas
25 x 35 cm

Sunlit Koi, Temperate House
watercolour
25 x 38 cm

Koi in the Temperate House
watercolour
28 x 37 cm

Koi with Reflections, Temperate House
watercolour
36 x 27 cm

Kew Green, Twilight
oil on canvas
25.5 x 76 cm

Kew Bridge, Autumn
oil on canvas
25.5 x 76 cm

The Lake, August
oil on canvas
30.5 x 38 cm

Morning Sun from Kew Bridge
oil on canvas
25 x 51 cm

Autumn Evening, Kew Gardens
watercolour
20.5 x 29 cm

The Queen's Garden
watercolour
22 x 30 cm

LIZ BUTTERWORTH

Owl, Kew Gardens
ink on paper
15 x 31 cm

Indian Ringneck with Chicks at Kew
pencil, ink, gouache on paper
51 x 30 cm

Autumn, The Cherry Walk
oil
63.5 x 76 cm

Summer Shadows
oil
63.5 x 76 cm

Frosty Morning
oil
76 x 50.5 cm

Palm House Visitors
watercolour on card
27.5 x 20 cm

CELIA CRAMPTON

Fallen Angel
Inspired by Emily Young's sculpture in Kew Gardens
watercolour on vellum
30 x 30 cm

In the Waterlily House
watercolour
24 x 26 cm

In the Temperate House
watercolour
29 x 20 cm

The Temperate House
watercolour
55 x 41 cm

The Palm House
watercolour
56 x 45 cm

The Pagoda from the Palm House
watercolour
29 x 44 cm

The Order Beds
oil
38 x 56 cm

Bamboo with the Minka House
watercolour
33.5 x 48.5 cm

Plane Leaves with Zita Elze Flower Shop
watercolour and pencil
30 x 16.5 cm

Cedar with the Palm House
watercolour and pencil
31.5 x 48 cm

Pond House, Kew Green
watercolour
46 x 51 cm

The Greyhound, Kew Green
watercolour
53 x 43 cm

Temple of Bellona
watercolour
27 x 18 cm

Bluebells
oil
35 x 46 cm

Early Spring
oil
38 x 46 cm

Promenade
watercolour
80 x 66 cm

Feeding the Geese
oil
74 x 91 cm

So Hard to Say
oil on canvas
56 x 46 cm

Nonchalant
oil on canvas
56 x 46 cm

Across the Thames to Syon House
oil on canvas
30 x 45 cm

Pagoda
oil on canvas
30 x 45 cm

Palm House
oil on canvas
30 x 45 cm

A Sunny Morning in Kew.
What's more alarming – planes every 3 minutes or the big flocks of screeching parakeets?
mixed media on paper
57 x 62 cm

Across the Thames from Kew to a Fox and Crows and Gulls
mixed media on paper
57.5 x 62.5 cm

Upstream - Sallow, mugwort, buddleia, hogwood, sycamore, burdock, plantain, ash, elder, dead nettle, cranesbill, ivy
mixed media on paper
57.5 x 61 cm

The Wedding Shop
oil on canvas
61 x 45 cm

Oliver's, Kew Village
oil on canvas
44 x 56 cm

A Front Garden, Kew
oil on canvas
61 x 76 cm

Moorhen at the Pond on Kew Green
watercolour and ink
42 x 57 cm

Golden Pheasant in the Redwood Forest, Kew Gardens
watercolour and ink
28 x 51 cm

Parakeets at Queen Charlotte's Cottage
watercolour and ink
31 x 19 cm

Guinea Fowl in Kew Gardens
monoprint and painting
65 x 48 cm

Golden Pheasant in the Redwood Forest
monoprint and painting
43 x 64 cm

Common Toad at the Lilypond, Kew Gardens
watercolour and ink
33 x 42 cm

CHRISTABEL KING

Camellia and the Temple of Bellona
watercolour and pencil
25.5 x 21.5 cm

Succulent
oil on canvas
46 x 56 cm

Rock
oil on canvas
46 x 56 cm

Hercules and Achelous, the Palm House
oil on canvas
71 x 107 cm

CARL LAUBIN

Victoria Regia House I
oil on canvas
30 x 25 cm

Victoria Regia House II
oil on canvas
71 x 183 cm

The Mediterranean Garden
oil on canvas
30 x 25 cm

Botanical Trophy
oil on canvas
137 x 102 cm

Golden Delicious
collage on board
122 x 74 cm

Night View, Kew Railway Bridge
watercolour
25.5 x 26.5 cm

Sun and Topiary,
Princess of Wales Conservatory
watercolour
25.5 x 26.5 cm

Sculling Under the Railway Bridge
oil
60 x 45 cm

Boats at Low Tide (near Kew Bridge)
oil
25 x 30 cm

Lasting Impression
watercolour and pencil
97 x 120 cm

Pagoda Tree
watercolour and pencil
48 x 39 cm

Autumn Afternoon
oil on panel
46.5 x 65.5 cm

The Queen's Garden
oil on panel
47 x 66.5 cm

St Anne's Church
oil
46 x 61 cm

Victoria Gate
oil
51 x 61 cm

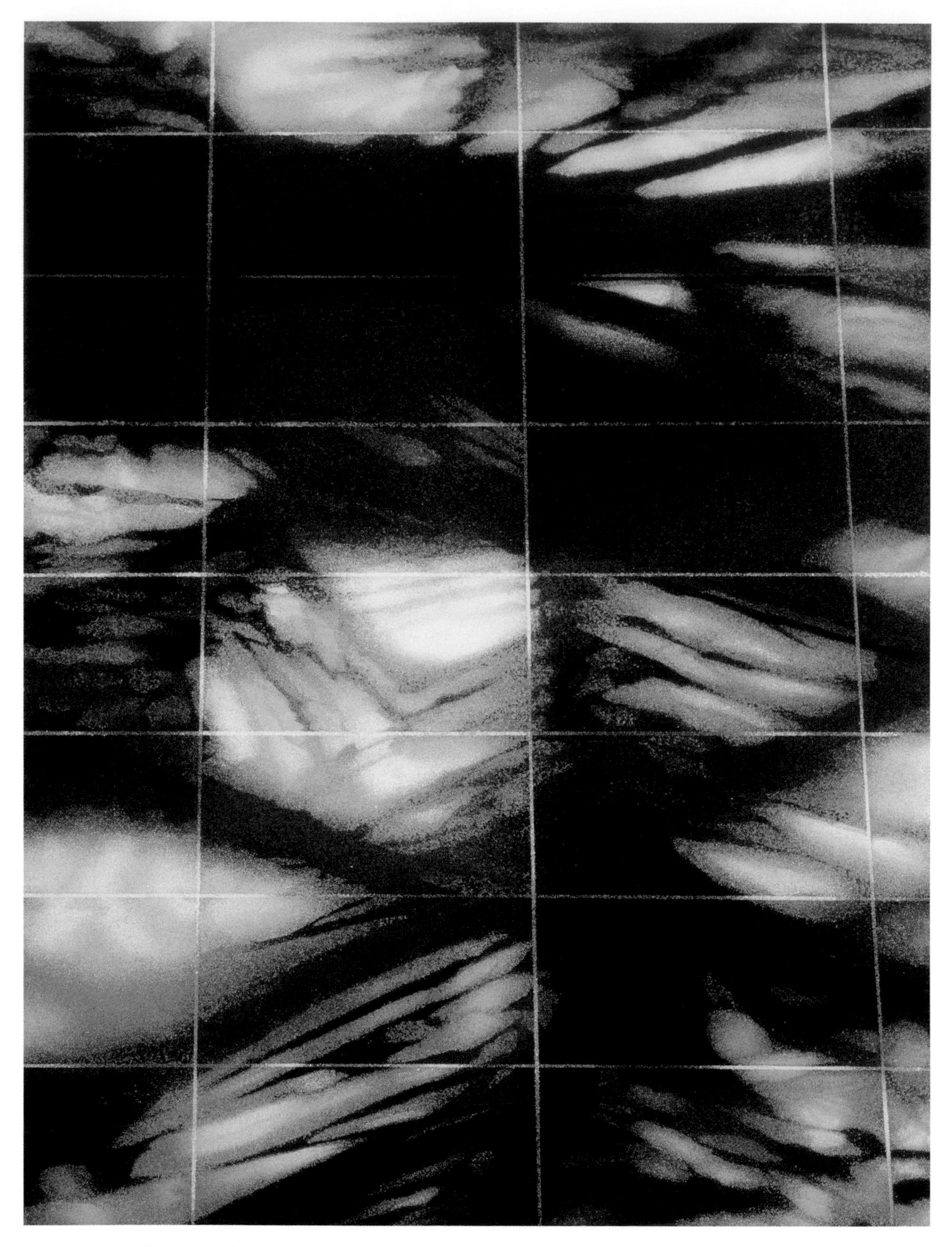

Palm House (Filtered Light)
acrylic on canvas
124 x 94.5 cm

Kew Pathway (Walkway)
acrylic on canvas
124 x 94.5 cm

Afternoon Sunlight
oil
30 x 40 cm

Kew Bridge
watercolour
25 x 35 cm

Kew Bridge and Mooring
oil on canvas
50.5 x 61 cm

Kew Gardens Pier
oil on canvas
20.5 x 35.5 cm

Jetty from Kew Bridge
watercolour
59 x 26 cm

Regimented Cacti
watercolour
71 x 22 cm

Cutting from Kew Gardens
ink on paper
50 x 71 cm

Holly Walk, Late Afternoon
oil on canvas
122 x 91 cm

View to Syon House
acrylic on board
55.5 x 75 cm

Encounter in Kew
pastel on canvas
76 x 61 cm

Kew Romance
pastel on canvas
76 x 51 cm

January Afternoon, Kew Green
oil on canvas
14 x 22 cm

PANDORA SELLARS

A Kew Griffin
gouache and pastel
8.5 x 11 cm

Richmond Hound
gouache and pastel
24 x 18 cm

JANET SKEA

Still-Life: Marianne North
watercolour
39 x 55 cm

Kew Still-Life
watercolour
94 x 77 cm

Cycad (Encephalartos altensteinii), Palm House
watercolour and gouache
41 x 28.5 cm

Palm House Gallery
watercolour and gouache
22.5 x 10.5 cm

Springtime
watercolour
29 x 48 cm

View Across the Lake at Kew Gardens
watercolour
29 x 41 cm

Laburnum Walk in Spring
watercolour
29 x 41 cm

Woodland Garden and Temple of Aeolus
watercolour
41 x 29 cm

The Palm House
watercolour
29 x 47 cm

The National Archives
oil
30.5 x 40.5 cm

Selecting the Specimen of Convolvulus
pastel and watercolour
80 x 51 cm

Study for Selecting the Specimen of Convolvulus
pastel and watercolour
48 x 28 cm

Preparation of the Alpine House
pencil and watercolour
38 x 58 cm

Football Practice, Kew Green
pencil and watercolour
38 x 58 cm

Temperate House
oil
56 x 152 cm

Student Vegetable Plot
oil
41 x 39 cm

Japanese Gateway
oil
31 x 36 cm

Autumn
oil
25 x 39 cm

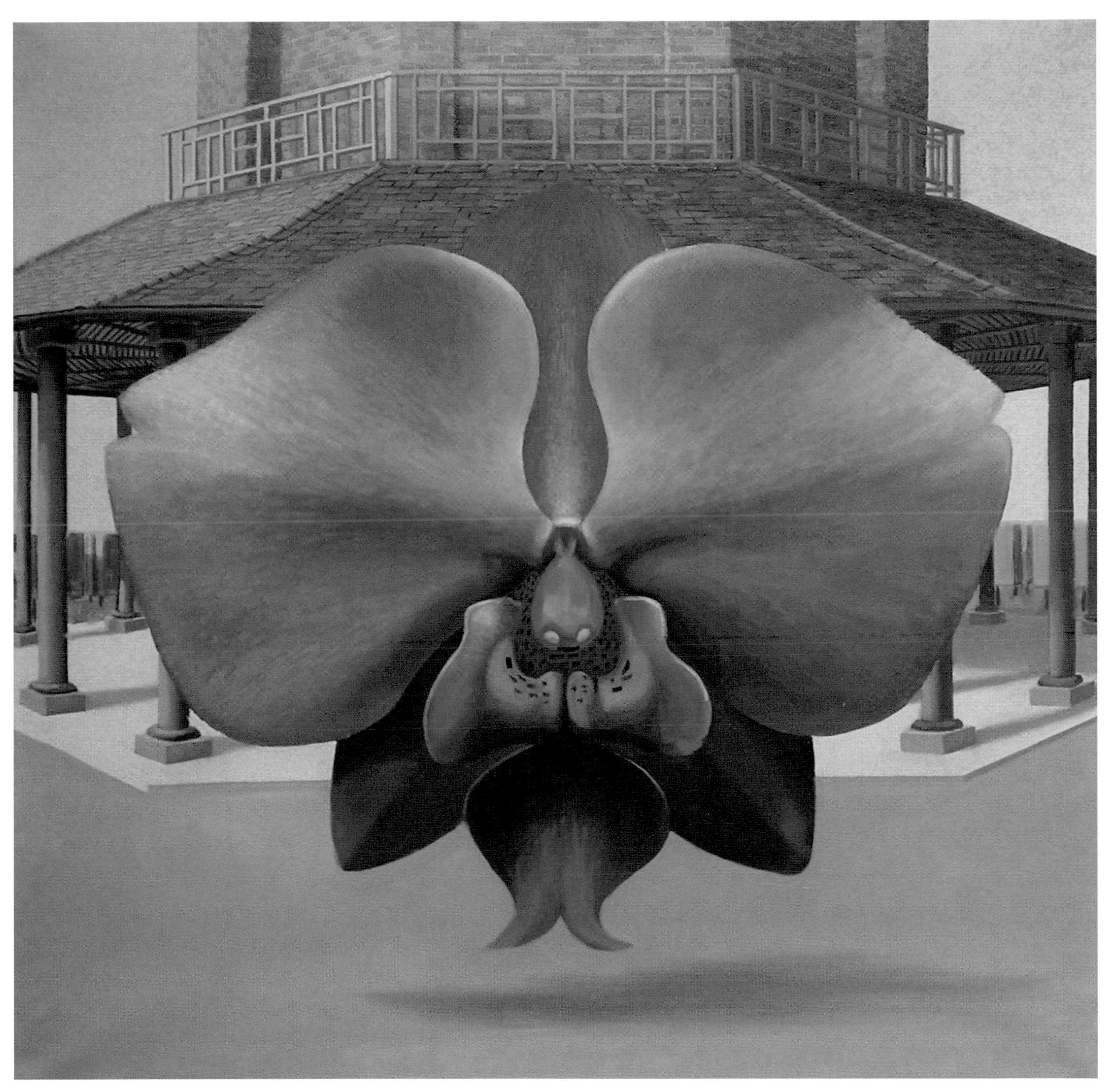

Pagoda Moth
oil on canvas
80 x 80 cm

Mirangerie
oil on canvas
61 x 122 cm

Dreaming of Victorias
oil on canvas
101 x 152 cm

Kew Bridge
ink
20 x 29 cm

Autumn, Kew Gardens
oil
23 x 28 cm

The National Archives
charcoal
55 x 76 cm

St Anne's Church, Kew Green
charcoal
42 x 42 cm

Sight Screen, Kew Green
charcoal
42 x 42 cm

The Artists

Matthew Alexander

Naomi Alexander ROI

Glenys Ambrus ARCA PS

Victor Ambrus FRSA RE ARCA PS

C. J. Archer RI

Michael Aubrey

Sir Peter Blake

Nick Botting

Francis Bowyer PPRWS NEAC

Peter Brown NEAC ROI

Liz Butler MA(RCA) RWS

Elizabeth Butterworth

Jane Corsellis NEAC RWA RWS

Celia Crampton

Claire Dalby RWS RE

John Doyle MBE PPRWS

Ann Farrer

Pauline Fazakerley

Olga Geoghegan

Angela Gladwell MA(RCA)

Peter Harrap

James Hart Dyke

Kurt Jackson

John James

Rebecca Jewell

Christabel King FSBA

Natasha Kissell

Carl Laubin

David Mach RA

Ronald Maddox PRI Hon.RWS Hon.RBA

Luke Martineau

Angus McEwan RSW

Sally Moore

Anthony Morris RP NEAC

Brendan Neiland

Denis Pannett G.Av.A.

David Paskett RWS

David Parfitt NEAC

Martha Richler (MARF)

Paul Roberts

Zsuzsi Roboz

Alex Russell-Flint

Pandora Sellars

Janet Skea RI

Lucy Smith

Sergei Sukonkin

John Ward CBE NEAC

Toby Ward

Steve Whitehead

Michael Whittlesea RWS NEAC

Antonia Williams

Robbie Wraith RP

Roy Wright

Acknowledgments

Matthew Alexander, Peter Brown, Jane Corsellis, Kurt Jackson and Zsuzsi Roboz by arrangement with Messum's

Nick Botting, Portland Gallery

Liz Butler, Francis Kyle Gallery

James Hart Dyke, John Mitchell Fine Paintings

Carl Laubin, Paul Roberts and Antonia Williams, Plus One Gallery

Luke Martineau and Steve Whitehead, Panter & Hall

Sally Moore, Martin Tinney Gallery (Cardiff)

Brendan Neiland, Redfern Gallery

Alex Russell-Flint and Robbie Wraith, Petley Fine Art

Federation of British Artists and Natural History Museum for general assistance.

Messum's for providing their gallery in Cork Street as the venue for a central London preview and for hosting a reception there.

Featherstone Leigh for logistic support in Richmond.

STEVE WHITEHEAD
Waterlily House
acrylic
119 x 53 cm

Epilogue

My first sight of the images appearing in this book was salutary. It reminded me that Kew is itself a work of art, shaped by Nature and by the vision and the hands of earlier generations, and with an eclectic richness that gives Kew its strong sense of identity.

At the Royal Botanic Gardens, our role is not simply to display the variety of plant life, but also to improve our collective understanding of the world of plants and our appreciation of Nature. Through the unrivalled collection of botanical art already held at Kew, and through exhibitions like the one celebrated in these pages, we can offer a new way for thousands of visitors each year to appreciate the creativity and beauty of Nature and perhaps also to see the Kew community in a new light. Thus, *Artists' Kew* is more than art for art's sake. It is an encouragement to engage with art and to reflect on what makes Kew special.

Funds raised from this exhibition will support plans to make our collection of art more accessible to the public. It is a project which Kew's Director, Sir Peter Crane, has championed. His energy and vision have enabled us to move towards the construction of a new gallery within the Gardens to display Kew's own artistic treasures. While these plans take shape, all of us look forward to the Gardens' 250th anniversary in 2009.

For the Royal Botanic Gardens, the next 250 years will bring ever greater challenges as we work to address the issues of habitat destruction and changing climates. Future generations should inherit an environment as rich and fascinating as that passed on to us. We shall continue to do all we can to influence those future generations to value their surroundings as highly as we do ourselves.

The Earl of Selborne
Chairman, Board of Trustees
Royal Botanic Gardens, Kew

February 2006

CELIA CRAMPTON
Secret Garden
Inspired by stained glass in St Anne's Church
coloured ink on card
16 x 14 cm